Gemba Walks for Service Excellence

The Step-by-Step Guide for
Identifying Service Delighters

Gemba Walks for Service Excellence

The Step-by-Step Guide for
Identifying Service Delighters

Robert Petruska

CRC Press
Taylor & Francis Group
Boca Raton   London   New York

CRC Press is an imprint of the
Taylor & Francis Group, an **informa** business

A PRODUCTIVITY PRESS BOOK

CRC Press
Taylor & Francis Group
6000 Broken Sound Parkway NW, Suite 300
Boca Raton, FL 33487-2742

Printed in the United States of America on acid-free paper
Version Date: 20120516

International Standard Book Number: 978-1-4398-8674-8 (Paperback)

Library of Congress Cataloging-in-Publication Data

Petruska, Robert.
   Gemba walks for service excellence : the step-by-step guide for identifying service delighters / Robert Petruska.
      p. cm.
   Includes bibliographical references and index.
   ISBN 978-1-4398-8674-8
   1. Customer services. 2. Organizational effectiveness. 3. Customer relations. I. Title.

HF5415.5.P486 2012
658.8'12--dc23   2012002430

Visit the Taylor & Francis Web site at
http://www.taylorandfrancis.com

and the CRC Press Web site at
http://www.crcpress.com

Contents

Supplementary Resources Disclaimer

Additional resources were previously made available for this title on CD. However, as CD has become a less accessible format, all resources have been moved to a more convenient online download option.

You can find these resources available here: www.routledge.com/9781439886748

Please note: Where this title mentions the associated disc, please use the downloadable resources instead.

Preface

It was an honor to be selected by the American Society for Quality (ASQ) to present at a Lean Six Sigma conference in Phoenix, Arizona. My presentation consisted of hand-drawn posters and large blown-up photographs that were hung around the entire conference room like an unveiling at an art exhibition. I was almost embarrassed with how much time and energy I had put into preparing for a one-hour presentation. Soon after starting my presentation, everyone was up out of their seats and walking from poster to poster while I told stories from experiences with service excellence.

About fifteen minutes into my presentation, the back door opened up and several new people entered the room to join us. I found out later that some participants must have really liked the presentation because they were broadcasting positive comments to their cloud of followers in real time. Positive tweets had attracted new arrivals to join in the fun. Serendipitously, a senior editor with Productivity Press, Michael Sinocchi, met me after the presentation, and he suggested that I write a book on this topic. Despite having a completely different background than most other conference participants, he could readily understand my presentation. Therefore, he thought this topic would appeal to others as well. The feedback from my presentation was excellent, and taking all of this into account, I decided to jump in with both feet. Readers will notice many hand-drawn illustrations, which I created myself. It is obvious that I am not an artist, but I hope my artwork will help convey meaning and resonate with you.

Thank you for buying this book, and for your efforts to improve service.

Acknowledgments

I am grateful to my wife, who allowed me to put her pictures in this book to help illustrate some key points. I also appreciate my daughter, who as a high school student helped correct my grammar along the way.

I am also appreciative of my many friends and mentors who helped me learn and grow.

Introduction

"Irasshaimase!" she said as I stepped inside. It was my first trip to Japan, and I had just arrived in Kobe during the heat of summer. I was starting to get that sinking feeling in the pit of my stomach that comes from the realization that something was massively out of place. It's also called "culture shock." I am ashamed to admit it, but my Japanese vocabulary was about three words at the time, and *irasshaimase* was not one of them.

The young woman wore an impeccably neat uniform. Her standing posture was perfect, her long dark hair held firmly in place, and she wore white gloves all the way up to her elbows. Her convenience store was spotless, and the store merchandise was arranged in such a manner that you could easily see everything displayed.

She quickly made eye contact with me and smiled while bowing gracefully to show deep respect. The meaning of her greeting had finally dawned on me; she was welcoming me into her store with sincere honor and gratitude. This was at a convenience store in an otherwise busy, drab, and crowded Japanese train station!

As I walked toward the back of the store, I felt strangely comfortable. I didn't feel as if I was the enemy—always under a watchful eye, another potential thief expected to pilfer merchandise at any moment. Instead, I felt more like a welcome guest in a good friend's home.

Soon, the attractive store worker walked away from the counter and gently approached. She spoke rather good English, which was a huge relief. With three different alphabets consisting of over 10,000 unique letters, Japanese was not a language I could master during the flight there. She helped me find what I was looking for despite a language barrier, then quickly rang up my purchases and sent me on my way.

My mental image of a typical convenience store had just been shattered. The chards of dissonance left behind pointed me toward exploring how some service providers could pleasantly surprise customers. I had never seen anyone in any store treat customers as nicely as she just did. Was this something cultural, or was this something else?

Ever since that day in Japan, I have been fascinated with service excellence. This book contains a collection of observations and experiences that are deconstructed and boiled down into their essence.

The purpose of this book is to spark the counterintuitive thinking needed to create fresh sources of customer delight. Customers are attracted to companies that are on the leading edge of service innovation, and will handsomely reward those who occupy this space. Service delighters eventually lose their appeal, and this erosion necessitates a concerted effort to create an idea pipeline full of innovation. Thinking of new ideas is just the first step, and serious competitors must also learn how to engage their staffs to experiment, measure, and broadly implement new ideas. Implementation is always the hard part of any effort to improve. The companion CD has innovative "placemats" designed to provide stepping stones on a development path for your team to create an increased competitive advantage in the marketplace (see Figure 0.1).

This book will benefit those readers who are looking for a source of new ideas to implement by sharing proven techniques used in Lean manufacturing that can be readily applied to the service industry. Readers will learn how to use Gemba walks to identify new service delighters, and to find out what their situation is firsthand. By deeply reflecting on experiences that can only come

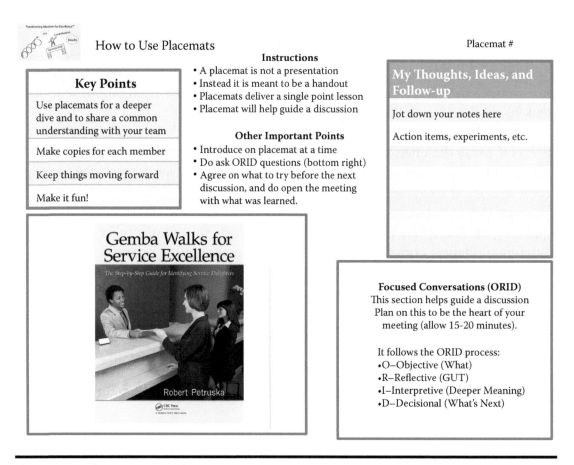

Figure 0.1   Placemat example.

from direct observation, readers will find themselves with a steady stream of new ideas and find ways to engage the staff with the entire process.

This book will function as a step-by-step guide for the reader to embark on a journey of finding and implementing new service delighters. Placemats are designed to be used by teams to provide structure for people to learn and reflect on new concepts. They are designed to be rolled out in sequence, and each one is a single point lesson designed to help people grow. It's best to take the time needed for people to digest each concept before moving on to the next one. It would also be beneficial to consider this effort as a continual operating system for service improvement rather than a single point improvement.

Readers are given a glimpse into the service experiences that have delighted customers. These experiences are dissected to their core components or essence. Illustrations are intended to stretch people's imaginations as they engage in a creative process of organizational growth. Each core service component is considered individually and from a system of work perspective. Placemats are best used in a consistent manner with teams of people who want to be part of the solution. It is always helpful to have a guide who can inspire the team to action, and this person will be instrumental in sustaining momentum over the long haul. Stretch assignments help provide a steady source of new ideas while developing and implementing a go-forward plan of action. Active participants will definitely find some quick wins from using the techniques in this book, and because your competitors are not standing still, ongoing efforts are needed.

So let's begin with a conversation around innovation. As you are undoubtedly aware, many organizations have made innovation a top strategic priority. But what exactly is meant by innovation? It seems we see examples of product innovation daily, and that begs us to ask the question, "Why doesn't service innovation seem to occur more frequently?" or "Why don't we seem to notice service innovation when it does occur?" Are we so accustomed to poor services that we attempt to blot out the experiences from our minds? For example, when our family relocated to North Carolina in 2011, the state required us to visit its governmental office to obtain our driver's licenses. It literally took five hours to get the coveted photo identification, and about 5 percent of that time was spent in value-added activity.

We all have horror stories about things that have gone wrong, and an entire book could easily be written about bad service. This book focuses on things gone right instead using a process similar to *positive deviance*, and this stems from observing service transactions directly with an eye for those rare *sparks of innovation*. A collection of experiences are shared with the reader to provide examples and case studies that we dissect together. As we dig in deeper and analyze what happens behind the scenes, a recipe for your organization to deliver excellent service will emerge. These observations will serve as a prompt for your entire team as you create your own blueprint for exploring, identifying, and experimenting with new service delighters on your way to improved business results at your own organization.

This workbook is intended to be used as an experiential learning guide as you begin exploring and experimenting with potential service delighters. It is critical to seek the thoughts, ideas, observations, and experiences of the front-line employee to get the most benefit from this book. The chances of new ideas being sustained increase dramatically if the people doing the work come up with the ideas themselves. Ideas that are suggested by an outside expert elsewhere in an organization naturally encounters resistance. Some of the best ideas come from people working on the front lines, and in many cases, others are just waiting on the sidelines for you (the coach) to put them on the playing field and pass them the ball. Therefore, the techniques in this book will be invaluable for understanding the essentials of service as well as applying them with your own team.

The next chapter will focus on leadership styles. While some traditional managers have evolved, others are operating in the same frame of mind that was common in the 1980s. Let's take a close look at what is meant by a *Lean leader*, and how this mindset might help engage people's heads, hearts, and hands much better than a traditional *great leader* model.

Chapter 1

Lean Leadership

The era of the great leader is over. It was built on a model that is no longer relevant. At one time, great leaders were expected to be the *experts*, and were entrusted to make nearly all of the decisions. Even early organizational improvement models involved a very few number of technical experts with deep technical knowledge, and then they would proceed to study a problem to death. They were expected to provide only the facts, and perform detailed analyses needed to thoroughly describe the cause-and-effect relationship. Our technical experts would literally collect reams of data, perform carefully designed experiments, display fancy charts with impressive statistics just to prove that change was needed. In some cases it would turn out to be a relatively simple change, one that could have been done weeks earlier. There is nothing wrong with studying problems using statistics, and in some situations, this is the best way to go forward. For the purposes of this book, the focus is on transforming leadership mindsets within the service industry. Therefore, if your organization is operating with the mindset that only a few experts know the answers, then there are a few things that you will need to tweak as you embark on your own Lean journey.

Self-awareness is a good starting point. One of my mentors introduced a concept that was rather interesting. He called it *unconsciously ignorant*, which describes a person who is ignorant of something and at the same time is oblivious to their own ignorance. What a fascinating concept! Unless someone who we trust sufficiently well has enough courage and permission to clue us in on our own ignorance, we'll continue to go down the path we're going without even being aware. We just don't know what we don't know!

The good news is self-awareness is a competency that each person can work on. To begin the journey as a Lean leader is to start with looking closely in the mirror to see what those around you are seeing. You might not see the image you wanted to project, and that's okay for now. All of us are working on improving ourselves. Otherwise, we wouldn't even begin this journey (see Figure 1.1).

Figure 1.1   The mirror never lies (or does it?).

Think of the great leader model as an elephant sitting on an organization's chest—stifling creativity. Great leaders were expected never to show any signs of weakness. If he did not know an answer to any question, he would just talk louder. People working in these kinds of environments tend to do exactly what they are told. Discretionary effort is not rewarded, and in some cases it's actually punished! Woe to the person who brings bad data to a meeting. Shoot the messenger! People with ideas for how things could be improved in these environments find themselves stymied.

People working in top-down organizations might have even tried to sell their ideas to their boss, but found a less than receptive audience. Sometimes their ideas were so good that they were quietly implemented later so the boss could take credit. Great leaders have to claim new ideas as their own to keep in the good graces of an even more powerful great leader.

Ultimately, managers became the bottlenecks for the flow of new ideas. It was the manager's prerogative whether something went forward or not. This environment leads to favoritism and pet projects appearing out of nowhere, which contributes to an atmosphere of suspicion and lack of trust. The impact of these types of behaviors typically results in a deep cynicism and helps create fodder for comic strips.

Lean leaders, on the other hand, recognize that they don't know all the answers, but they know which questions to ask to get the best out of their people. Lean leaders show respect by listening, and instead of being a bottleneck

Table 1.1   Lean Leadership Mindset

Creates a blameless environment
Knows the right question to ask, and when to ask it
Provides ample opportunities for recognition
Encourages experimentation as a way to prove ideas
Respects people
Good listener and makes time for employees
Regularly shares customer feedback
Sets the bar high
Develops talent through coaching and mentoring
Energizes and influences people
Walks the talk, has high integrity
Interested in how things actually get done
Gets into the messy details by checking
Prevents problems
Thinks systemically; understands that the system of work is their responsibility
Encourages involvement
Sees value in different perspectives
Rewards efforts as well as results
Learns best by doing
Builds relationships from trust
Likes to go and see for themselves

for implementing new ideas and thoughts, they cultivate a well of ideas, and can pull from that well as needed (see Table 1.1).

In summary, the world has changed, and we can no longer assume that one person knows it all. We've evolved into organizations of interdependent people, and depending where you are in an organization, you may have to take a different tactic to begin.

Formal Leader

If you're in the position of a formal leader, you have a great advantage to getting started with any improvement effort. The people working for you naturally want to complete tasks that result in some kind of reward. For example, your active presence in a meeting or the manner in which a team member is publicly shown respect can be very rewarding to those who work for you.

As a formal leader, some of your most important tasks are to grow talent, develop capabilities, and unleash creativity. The good news is you can do all three by mastering just one skill—*delegation*!

While consulting in the healthcare field, I was surprised by how overburdened some of the managers said they were. At the root of their work–life balance issues, we found that some healthcare managers were actually not comfortable with delegating. It also seemed that delegation was a skill that some managers really hadn't mastered at that time. As it turns out, delegation was one of the ways in which the manager could achieve a better work–life balance. As we dug deeper, a deep ingrained fear of failure was uncovered. Failure would make any manager look bad to their peers or boss. So there was a kind of vicious circle going on; managers uncomfortable with delegating, therefore people were not given stretch assignments, resulting in lack of growth in capability and career stagnation (which likely contributes to turnover). Ultimately, the manager is stuck doing it all, which causes the work–life imbalance to begin with.

Delegation is also liberating. Think of your child learning to ride a bicycle for the first time without training wheels. You want your child to be able to keep up with you, but the training wheels keep holding him or her back. Only by removing the training wheels will they be able to keep up with the family on an afternoon bike ride. However, removing the training wheels results in your child wobbling and falling if you are not there to steady them as needed. Eventually, you have to learn to let go (possibly because you cannot run fast enough to keep up). At that precise moment in time, they have to learn to balance their body on two wheels on their own. No one ever learned to ride a bicycle by reading a book! You have to try it, fail, pick yourself up, and try again.

Getting Started

To engage people's heads, hearts, and hands, you've got to make things fun and rewarding. My suggestion is to make *fun* one of the ground rules for all of your meetings. Make sure people know this does not mean *make fun of someone else*. Interpreting and enforcing ground rules is also very important to running productive meetings. If your meetings are not normally described as fun, and you've seen any of these dysfunctional behaviors—closed body language, eyes rolling, frequent multitasking, or sidebar conversations—then remedial action is needed. If this is the case, you might want to consider asking someone who has the requisite skills to run your meeting as a neutral facilitator.

To Find Out Where You Stand

Try this experiment: simply invite your staff to a meeting with the caveat that no one is required to come. Only those who care enough about the meeting topic will attend. How many times have you been in a meeting where a few people, who probably shouldn't have been there, wasted everyone's time? It takes just one screwball to derail an entire meeting by taking everyone off topic or sending attendees down a blind alley. If a person doesn't want to be in a meeting, they shouldn't be there. So make your meetings voluntary!

You're probably wondering what happens if no one shows up at all. That is the risk you must be willing to take, and if no one shows up to the meeting it sends a strong message. The good news is that you're no longer unconsciously ignorant! It just means that you've got some work to do before people will be willing to be led in a natural, authentic way.

The goal of the first meeting is to introduce the concept of change and begin a gradual transfer of power. If you are comfortable with sharing knowledge, power, and successes with your team, then you're well on your way to becoming a Lean leader.

Let's start by talking about how a service differs from a product. *Service*, in its most basic definition, is a kind of human transaction where value is delivered from one person to another. A product is much more tangible; it's something you can touch. There was a time when our economy was driven primarily by manufacturing products. However, the U.S. economy is now primarily driven by services. Even if your organization is a manufacturer, service is an important component for your customers. How people are treated after a sale can be a key driver for repeat business. Even the suppliers who provide your organization with raw materials or third-party logistics interact with your employees on a human level, and can contribute to your brand's value proposition (both positively and otherwise). How suppliers and other key stakeholders are treated matters!

Explore with your team (see Placemat 1.1): Given the previous broad definition of service, create a list of all the different services your organization provides. Who are the customers for each service? Then take a look at the list. Is it longer than you would have thought? Have you identified any customers for the work you do yourself? What surprised you?

Explore with your team (see Placemat 1.2): Brainstorming is a key skill needed for any team to create new ideas to try, and this placemat is meant to provide a basic starting point so that the team gets comfortable working together.

Now let's take a fresh look at Professor Kano's model of customer satisfaction.

As you can see in Figure 1.2, the Y axis is the degree of customer satisfaction, and the X axis is the degree to which the service is being fulfilled. Professor Kano studied customer satisfaction and identified three different and distinct characteristics. The Kano model has been applied successfully to product design for many years, and is an essential key for companies to innovate with respect to their services. The first service characteristic is basic, which is really a price of admission or a must-have, without which a customer would be extremely dissatisfied.

For example, when I walk into a hotel lobby to check into my room, I expect to have the room that I reserved available, and if it's not available,

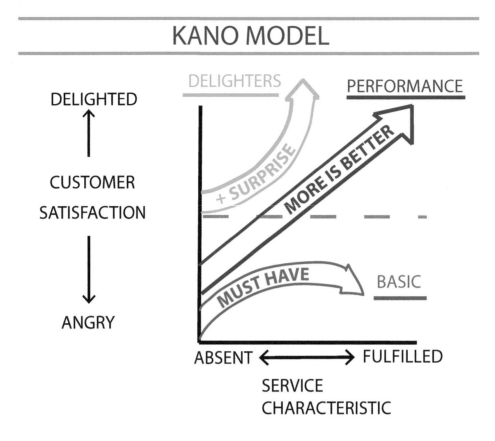

SEEK DELIGHTERS

Figure 1.2 The Kano model. (Adapted from Noriaki Kano, Nobuhiku Seraku, Fumio Takahashi, and Shinichi Tsuji, "Attractive Quality and Must-Be Quality" [in Japanese]. *Journal of the Japanese Society for Quality Control* 14, no. 2 [April 1984]: 39–48.)

I would be extremely dissatisfied. However, if the hotel has 100 open rooms available, it doesn't necessarily add to my satisfaction level. So a basic characteristic will not increase customer satisfaction with an increased presence. It just needs to be there. In fact, you could say that if you had too much of a basic characteristic, it might even degrade the experience. For example, if there were five clerks milling around at the front desk, I might begin to wonder if the management of the hotel is wasting money by having too many clerks.

Performance characteristics are service characteristics that Kano observed to behave linearly, having a positive slope. More is better. Therefore, if a service provider delivers an increased presence of a performance characteristic, the resulting customer satisfaction will increase commensurately.

Many service providers focus energy on improving performance characteristics within their operation and that will lead to a competitive advantage in the marketplace. While it is true that customer satisfaction will increase, fulfilling performance characteristics comes at a steep cost because the more effort you put in, the more costly it is to your organization. For example, think of a server at a restaurant. Your server has an uncanny sense of timing. She seems to be able to waltz into your dinner party without interrupting anyone. She somehow gets her job done, but doesn't interrupt the conversation. She actually adds to your party's overall dining experience. Your server arrives just before anyone thinks to ask to fill your drinks. Your food is delivered piping hot, and everyone's order is correct. The evening is going perfectly. She arrives with the check at just the right moment in time, and instinctively senses who is picking up the tab. Think about what it would take for any organization to recruit and develop such a key employee. To what lengths might that organization have to go to keep such a valuable asset happily employed?

Increasing performance characteristics comes at a steep price! That price is the quality of the person as well as their training and their compensation. If you are going to increase customer satisfaction solely through performance characteristics, you are going to just have to work harder or pay more. Ultimately, focusing solely on increasing performance characteristics may not deliver a big enough return to provide a sustainable competitive advantage over the long haul.

Kano's service *delighter* characteristics are new to the world, and they are always a surprise. They are something innovative, and provide a lift for the customer. It's a pleasant surprise to the upside. One way to test whether a service characteristic is a delighter is to ask your customer, "Would you mind if that characteristic were not present?" If a customer wouldn't mind if a particular characteristic was not present, and if the customer tells you they would be happy if the same service characteristic were delivered, it is most likely a delighter.

Generating a steady stream of new service delighters is one key to providing a distinct competitive advantage to companies in any service industry. And guess what—the good news is that finding new delighters isn't as hard as you might think! In fact, it isn't as hard as increasing customer satisfaction by grinding out your table stakes (performance characteristics). This is because the Kano model shows a very nonlinear relationship between delighters and customer satisfaction, which means even a small presence of a service delighter goes a long way toward delighting and exciting your customers.

Central to the approach outlined in this book is the concept of *portability* and *transference*. Service is quite universal in nature due to the human element, such that a service innovation in one industry could readily transfer to another. You may have to tweak service characteristics a bit for your unique situation, but there is no reason a common practice in the hotel and hospitality industry couldn't be tested out and adopted in the healthcare arena. Most importantly, this could be a new delighter for the healthcare business, and many of the new service delighters can be implemented with relatively low cost!

Because customer expectations tend to always increase, service characteristics actually erode over time. Erosion occurs all too often even as companies create a new service delighter, only to see in a relatively short time that it becomes a performance characteristic, which later degrades into a basic characteristic.

Figure 1.3   Pager/coaster.

For example, at the restaurant at one time, a service delighter would've been the hostess providing you with a drink coaster that suddenly flashes and turns into a UFO-looking thing when your table is ready (see Figure 1.3). The coaster or pager is an improvement over the previous manual process of shouting out "Jones, party of seven!" Over time, even that service innovation becomes expected. A delighter's novelty wears off as more and more people have encountered a similar delighter at competing businesses. When any innovation that was once considered a delighter becomes an expectation, it is not long before it erodes all the way down to a basic characteristic. To illustrate the concept of erosion, are you still delighted when an agent at the airline points you to a self-service check-in kiosk at the airport?

Your competitors are not standing still and may even be copying your innovations. This accelerates the rate of erosion and diminishes the value of your new service innovations. So how can anyone get to the top and stay ahead? Part of the answer lies in the heart of implementation; it is easy to copy the tools, but not the underlying principles. Another part of the answer is in applying a very cool technique we borrow from companies such as Toyota and it's called the *Gemba walk*. In Chapter 2 we provide you with context around a method to create a steady stream of new service delighters—Gemba walks.

Explore with your team (See Placemat 1.3): Let's brainstorm a list of delighter service characteristics in any industry.

Let's brainstorm a list of delighter service characteristics in any industry that have eroded, and let's talk about how long it took. Then let's review the lists. Were there any basic, performance, or delighter service characteristics from any industry that could be applied to ours? Would any of those characteristics get a *wow* from our customers?

Table 1.2   Prioritization Matrix

Idea	Impact	Effort
List all of the ideas and rank them.		
High Impact, Low Effort	High Impact, High Effort	
Low Impact, Low Effort	Low Impact, High Effort	

Now would be a good time to review a simple prioritization matrix. The basic concept is to visually place the ideas that could provide a *wow* for your customers on a sheet similar to the graphic in Table 1.2. When prioritizing ideas, it's helpful to look at the impact of a suggestion versus the complexity. High-impact, low-complexity ideas should be the first to get done. We typically divide ideas up into three different categories: *Just Do Its* (JDIs), *Kaizen Events*, and *Projects*. For the purposes of the first pass through this book, it is recommended that you spend time on priorities that have high impact and are not overly complex to implement.

Are there any ideas that fall into the priority zone that you could put at the top of the list?

Chapter 2

Gemba Walks

Gemba means precious place of work; it's the place of action where work gets done. Gemba walks are a very powerful avenue for leaders to showcase humble and appreciative leadership behaviors. A *Gemba walk* is a method to engage the workforce in their native environment. It could even be called *leadership anthropology.* At its simplest, a Gemba walk is like a plant tour but with a specific learning purpose and consistent frequency. In the medical profession, a Gemba walk is essentially *rounding,* where a caregiver (physician, nurse, clergy, or technician) walks to the patients (customers) and finds out firsthand what is going on. This is the key difference—a Gemba walk requires the presence of a leader(s) to observe directly. At a hospital, rounding is done at defined intervals depending on the care needed.

What you pay attention to as a leader absolutely matters! It matters to the people around you, and by consciously choosing to make a consistent effort to observe directly, you will learn a great deal about the current situation. I've seen entire cultures in organizations changed as a result of the effective use of Gemba walks. In many organizations, information is filtered as it's transferred up the chain of command to the boss. The higher you go in organizations, the more removed people can get from the current reality, and the more truncated information gets. This causes a significant disconnection between the top managers and the people on the front lines.

Win the battle or win the war? To illustrate this point, in one location when the senior leaders would take a tour in a manufacturing plant in Minnesota (USA), the area location managers would work really hard to whip the plant into shape. When the senior managers and leaders arrived, the plant was standing tall. It was cleaner and more organized than ever, and all of the visual production charts were up to date. As the senior leaders walked through, they were quite impressed with what a great job everyone at that location was doing. As they met different people along the tour route, they heard nothing but glowing comments about how well everything was running. The performance of the plant, at least for the period of time when senior managers were focusing attention on it, was outstanding.

So why then is it a mystery when the plant manager asks the same senior leaders for help, the request is not always well received? In this case, getting money approved and allocated to improve performance was nearly impossible. Organizational inertia crept in, armed with its insidious spike strips. Checks and balances were put in place by finance professionals who were primarily held accountable for fiduciary control and not performance improvement. The net effect was the capital equipment expenditures justification process itself prevented that facility from being able to invest. As a result, performance suffered. Reflecting on this disparity, one awesome leader realized that the picture we were painting for the senior leaders was distorted, and that distortion resulted in the perception that investment was not required. If you think about it, this makes sense from a senior leader's perspective. They just witnessed firsthand information, and saw everything was running great. So the decision to invest in a different location was made with the limited capital budget.

To address this limitation, it was decided that future tours would include pointing out some of the major system constraints to the senior leaders in addition to demonstrating high performance and potential. In this way, a more balanced reality was shared, and better decisions regarding future investments in infrastructure could be made.

Think about a time you had the opportunity to take a bigwig on a tour. Was the focus on things gone right? If so, it is natural. Everyone wants senior leaders to keep the facility's doors open. Having a balanced perspective is also important. The challenge is, how can your location be portrayed realistically so that the needed resources can be obtained while not looking so pitiful that someone closes the doors? The key to this is to have regularly scheduled Gemba walks (see Figure 2.1).

While I worked in the automobile industry, I was expected to solve our customers' problems on very short notice. Whenever there were urgent concerns about our products, I was one of the few employees who were sent to a customer's location at two in the morning. For background purposes, I worked in the tire industry during a time of heightened awareness about tire quality. I had to investigate the reasons for any customer complaints. If there were actual defects escaping our manufacturing facility in Charlotte, North Carolina, it was my responsibility to understand the cause or causes, suggest countermeasures, and prevent recurrences. Our newest customer had very high quality expectations, and it required multiple visits as we worked out ways to meet new requirements.

Standing in the Circle: A Fundamental Lean Principle

When I arrived in Georgetown, Kentucky, for my first time, I was asked to stand in one location on the tire mounting line for several hours at a time.

GEMBA WALKS

● **GO AND SEE FOR YOURSELF**

● **STANDING IN THE CIRCLE**

● **OBSERVE DIRECTLY**

Figure 2.1 Gemba walks.

My job was to inspect every tire before it was put on the vehicles. Our customer had a requirement for suppliers with quality problems; they had to have six consecutive shifts of production with zero defects. As you can imagine, it's extremely boring to spend an entire day in one spot just standing and watching tire-wheel assemblies being inflated. I hadn't realized it at that time, but I was being trained on a very powerful technique, and it's called *standing in the circle.*

By staying in one spot for a very long period of time observing, listening, and training all your senses on one repetitive process, you develop a very deep understanding. An expectation from standing in the circle was that you could intelligently recommend suggestions for improvement. This training also had a carryover effect that I brought back to my employer, and I used this method to solve problems on a weekly basis. The ability to make improvements is a core competency for Lean, and *standing in the circle* is one of the most powerful techniques.

The same technique used in manufacturing applies very well to service industries. By taking Gemba walks through any service organization from the customer's point of view, you can quickly learn what's really going on. If you practice standing in the circle while taking a Gemba walk, it will be possible to identify new service delighters. To illustrate this technique, let's go on a Gemba walk right now.

Gemba Walk: Design for the Customer

Let's review some pictures to help us get started on our first Gemba walk. Imagine you're staying at a business-class hotel that costs less than $100 per night (see Figure 2.2). Based on the price, you have some idea of what to expect. Let's walk through this hotel check-in process together and see how well our expectations are met. One of the essential items of any service transaction is to receive artifacts. Artifacts are something tangible that customers are given that supports the process of delivering service. Artifacts can be receipts, coupons, branded giveaways, menus, wine lists; in this case, it's the key to your room along with a paper sleeve that may be the only written evidence of your actual room number.

One thing that pops out at me from our first picture is the sign that helps orient the customer, "Welcome to Gulfport." The notable thing that reaches out at me is the warmth of a welcoming smile from the front desk clerk. A smile is

Figure 2.2   Hotel check-in. (With permission of Hampton Hotels Hilton Worldwide.)

Figure 2.3   Smile-meter by Omron uses sensors to capture a smile and grade it from zero to 100.

a universal symbol that has the same meaning regardless of language or culture (see Figure 2.3). Omron Inc. makes software to compare the desk clerk's face to over a million faces and gives a score depending on eye movement, the curve of her lips, and other aspects of her face. She needs a minimum score to "pass" and to face the customer. In a Tokyo subway, the warmth of a smile of workers is important enough to institute these devices. Could a smiling face help your company project a desired image? Besides investing in software, are there other ways to get the job done?

Notice the well marked exit signs as we walked down the hallway (see Figure 2.4). The exit signs (see Figures 2.5 and 2.6) are even marked in reflective tape located at the foot level. At first you might think this is overkill, however, looking at it more closely, what I realize is that someone cared enough about their customers and employees that they took the time and energy to make an exit pathway accessible even in the unlikely case that heavy smoke occurs. This automatically gives a favorable impression of the hotel, indicating that they really care about customer safety, which demonstrates a clear respect for people.

Signs (see Figures 2.7 and 2.8) are strikingly clear because the chosen colors contrast nicely with the walls. Braille has also been added to help orient any visually impaired guests. This is surely an added cost and this extra effort continues to build on a positive first impression we have so far.

Figure 2.4 Hallway view.

Figure 2.5 Exit sign with arrows directing people to safety.

Figure 2.6   Reflective exit sign at foot level.

Figure 2.7   No mystery doors here!

Figure 2.8   Signage with beautifully contrasting colors.

As we continue deeper into the hotel, we find ourselves taking the emergency exit stairwell (see Figure 2.9). On the surface we might not think much about it, but this one just happens to be painted, clean, and carpeted. Amazingly it is also air-conditioned! The stairwell (see Figure 2.10) is designed to be wide enough to accommodate a potential rush of people who may be in a hurry to exit the premises. There is no evidence of tobacco products or cigarette smoke from guests or staff taking a quick smoke break. It's refreshing to see evidence that someone anticipated that customers might want to climb a flight or two of stairs after being cooped up traveling all day.

Arriving at our room, we have another clearly contrasted sign. The room number (Figure 2.11) is very legible and even has Braille. This recurring theme of *welcome* reinforces the meaning of hospitality, and it starts to make us believe that we really are welcome. Notice that there is a unique mnemonic image that invokes ephemeral memories of our youth through some strikingly nostalgic albeit queer Americana. Memories and dreams can be very visual by nature, and our service provider realized that customers will have a better chance to remember their own location if they are given such an image. Presumably to make guests safer, we seldom see traditional metallic room keys with our room numbers etched into them. This change could have the unintended consequence of a customer forgetting their room number (at least it has happened to me).

Figure 2.9   Exit stairs.

Figure 2.10   Emergency exit stairwell.

Figure 2.11 Room number with unique picture.

Figure 2.12 Burglar proofing.

It's easy to notice that an additional layer of security has been added. A metal plate was added to inhibit a potential intruder from prying open our door while we inhabit the room (see Figure 2.12). The message is clear; our hosts care about our safety, security, and well-being.

What did your customer experience just before arriving at your business? Travelers may have not had a very pleasant trip and often go through a seemingly horrendous ordeal just to arrive at a hotel. If a customer traveled by air, she may have gone through a series of somewhat degrading security checks. Travelers are routinely treated like cattle, as security officials usher people through a maze like chutes that wind around in a labyrinth of long lines. Many times, travelers wait for a long time, and may have to show their ID multiple times just so they can get to their gate and then wait some more (see Figure 2.13). By the time a customer finally arrives at the hotel, they are likely discombobulated.

The traveler might have arrived in a city where he or she has never stayed before. Again, it's refreshing to see that someone has taken the mindset of a customer and considered how fatigued they might be. Arriving at a hotel room is a significant milestone for the customer, and is a moment full of anticipation for a well-deserved respite. How disappointing would it be if your key didn't work the first time?

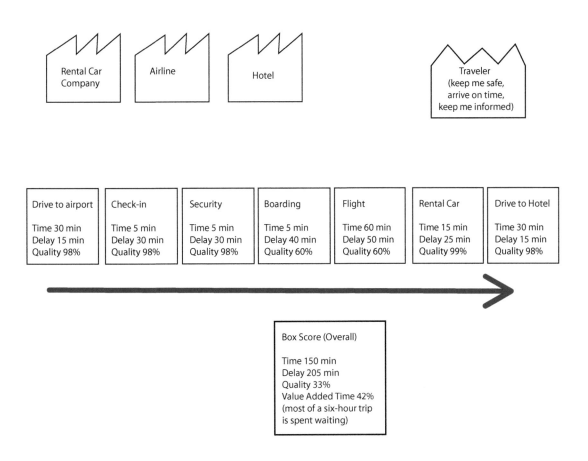

Figure 2.13   Travel value stream map (most of the time is spent waiting).

Figure 2.14   Key is inserted incorrectly in this picture.

This credit card style key is mistake-proofed by design, and there is only one way to insert the key to open your door (see Figures 2.14 and 2.15). Making the key easier for customers to use takes a little extra thought, and it pays off big time. Disoriented people can now get into their room easier because someone took the time to design the key to reduce aggravation. Imagine the frustration you might have if the key were designed with the same length and width. There could potentially be eight different ways you could use the key to open your door but only one of them would work. The design depicted cuts the number of possible ways to insert it in half because you can no longer slide the key in horizontally. By putting a picture on the front of this key, the possibility of insertion errors are cut in half again. This is the essence of mistake-proofing, intentionally making it difficult for us humans to err.

Combining the key holder with instructions on using the Internet (Figure 2.16), or how to check out conveniently at one time, was a nice surprise. As more hotels caught on to this innovation, we find artifacts that reinforce the brand's value proposition, such as these forget-me-not flower seeds (Figure 2.17) given as a gift and to celebrate the hotel achieving Green Seal Certified Hotel status. I look forward to planting these seeds in my garden, and when I enjoy the resulting flowers, I will remember this hotel (for a long time). Forget-me-not is a brilliant choice of flower name to give out as a gift. Imagine the thought that went into selecting this artifact?

Figure 2.15 Mistake proofing the key is easy as designed.

The hotel has also mistake-proofed the Do Not Disturb sign (Figure 2.18). In the past, Do Not Disturb signs would typically be something that fits over the doorknob.

Imagine if you happen to be traveling with a significant other, and in your haste to keep anyone from interrupting a romantic moment, you hurriedly put the sign on the door. But for whatever reason, the side facing the hall signals to the cleaning staff to clean up posthaste. Your amorous high point is brought to an abrupt end when an innocent staff member knocks on the door to clean up. By designing the sign to be identical on both sides (Figure 2.19) it now has only one meaning regardless of the orientation.

This hotel caters to business travelers who likely are under quite a bit of pressure. Hotel guests may be traveling to meet with one of their customers, suppliers, or even one of their satellite locations. As a result, hotel guests are constantly thinking. Creativity doesn't occur according to any schedule. Great ideas pop into our brains suddenly and sometimes without warning. It's as if our subconscious is constantly working to help us solve our problems or to dream up solutions. Innovation seems to occur when we least expect it, and when it washes over us like waves crashing on the beach, we may have only a few seconds before our once-great ideas wash out to the sea with the tide.

Notice the "thought pad" with a pen connected to it (Figure 2.20). Brilliant! This room had multiple thought pads located in various places, including the bathroom, desk, and even on the nightstand.

Figure 2.16 Key holder with an Internet access code.

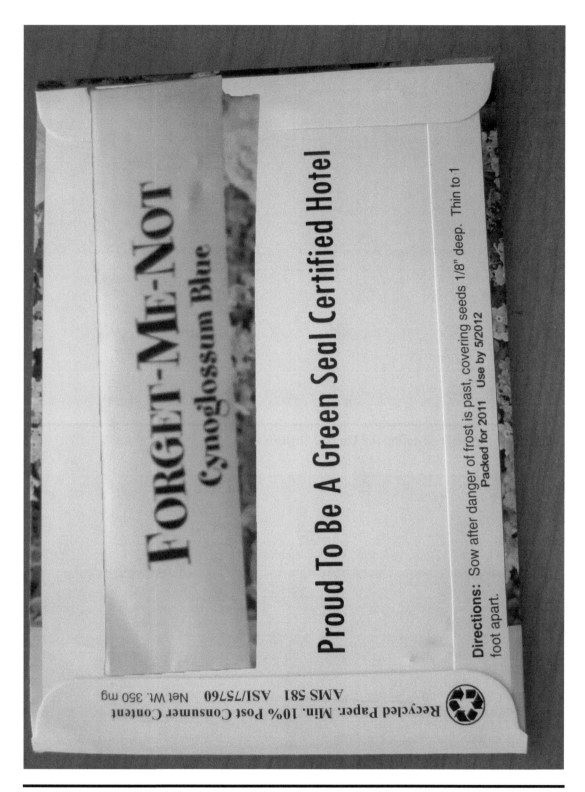

Figure 2.17 Nice artifact—a small gift with a hook.

Figure 2.18   Mistake proofing the Do Not Disturb sign.

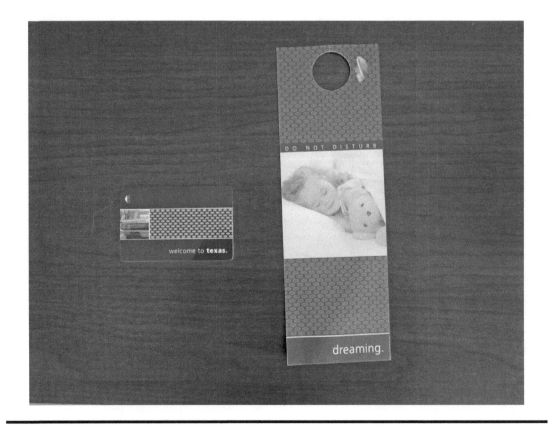

Figure 2.19   Welcome to Texas, and Dreaming.

Figure 2.20　Thought pad at bedside shows consideration for the business traveler.

You will also notice a very professional looking survey aptly titled "Fill Us In" (Figures 2.21 and 2.22). Impressively, this artifact is presented in a very meticulous manner. Clearly this is no ordinary feedback survey; it looks more akin to a birthday card. Whoever designed this format understood the value of customer feedback, and thought of ways to encourage customers to provide input to help them grow.

However, if the room were in a complete disarray—the bathroom had not been cleaned well, the sheets left unchanged, and the staff were unfriendly—we might conclude that filling it out would be a complete waste of our time. One might ask, "Why bother providing feedback to someone who obviously hasn't used any of the feedback they've received in the last ten years?" Providing feedback actually takes valuable time and energy away from customers. In this case, it's clear that this hotel organization is listening to their customers, and there is a very good chance that the staff would listen to any feedback offered.

In an age dominated by electronic communications, handwritten notes on very nice cards from housekeeping draw our attention to the fact that someone cares about us on a human level (Figures 2.23 and 2.24).

Entering the bathroom, I was impressed with how organized the shampoos and conditioners were (Figure 2.25). There were small pictures on each bottle accompanied by pleasant thoughts. *Purity* reminded me of the old saying "Cleanliness is next to Godliness." Thinking about it deeper, I imagine some people might have been traveling for close to 10 to 12 hours at this point.

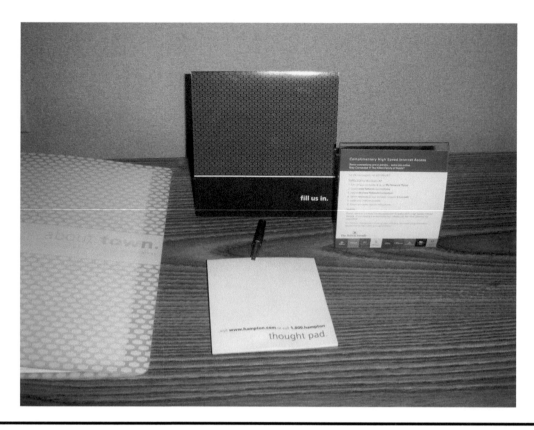

Figure 2.21   Fill Us In is a great way to invite customers to help.

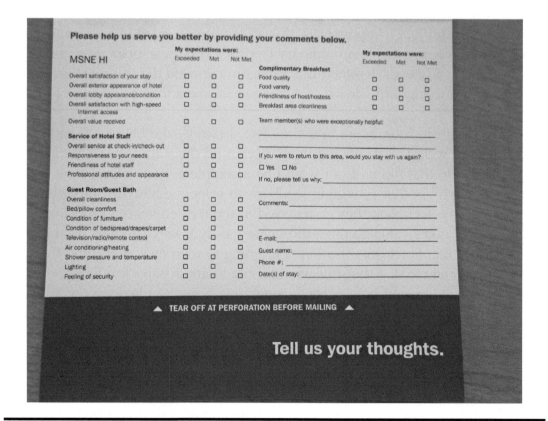

Figure 2.22   Customer feedback survey.

Figure 2.23 Thank-you note with a ribbon invites you to open it.

Figure 2.24 Authentic and caring, a handwritten note means a lot!

Figure 2.25 Purity gets to the point and resonates with how we feel at check-in.

U.S. domestic air travel has degenerated to the point where it's roughly analogous to riding a bus. The journey itself is uncomfortable, and by the time we arrive at our hotel rooms we feel rather unclean and we can hardly wait to take that shower or hot bath. Baggage fees discourage flyers from checking luggage, which results in more belongings carried onboard the aircraft. Increased security requirements have further reduced the amount of liquids travelers are allowed to carry, and some favorite soaps and shampoos inevitably get left behind. Soap that says *purity* on it invites us to take a moment to wash our hands and face and feel better right away. Even if we're too busy to take that shower or bath at that precise moment, we can take a few steps to clean up, calm down, and relax.

We may not be as comfortable as we are when we are at home, but at least this makes us feel better. Some hotels have even partnered with high-end bathing product manufacturers to provide small portions of very desirable brands to fill the void left by increased security mandates.

What if your spouse chooses your hotel based on the brand of bath products that she really wants for herself? Could this be a potential delighter? The hotel has to purchase bath products anyway, and the bath product supplier wants to reach new customers. What cool new ideas do you have for your own service based on this (See Figures 2.26 and 2.27)?

As we walk into the breakfast area, there is another sign with the name of the hostess working in the area (Figure 2.28). Introducing her by name demonstrates respect for people, and helps build pride in workmanship. If my name is on a

Figure 2.26 Branding and passive income opportunities.

board, I am going to do an even better job due to the influencing principle of *consistency*; people generally do what they have publicly committed to do.

Look at the thought put into designing the area where a customer might take a break and unwind (Figure 2.29). The space has been transformed from a plain room with four walls and a coffee machine into something very inviting and engaging. Suddenly I can imagine myself spending a Sunday afternoon with friends watching football, snacking, and cracking open a cold beer. And that takes my mind off of the fact that I am very far from home.

The breakfast area was surreal (Figure 2.30). Not only were sumptuous foods displayed, there were very clear signs and symbols helping the hotel guest to choose which food they might want to eat. There were universal symbols on those pictures that would help a visitor not familiar with the local language find what they wanted. Notice pictures of cows in the background, which are universal symbols of milk (Figure 2.31). The use of colors was also very appropriate and the deep blue background in this case made it clear these were cold items. See Figures 2.32–2.35 for other images of this breakfast display.

The "Catch Me at My Best" survey (Figure 2.36) focuses entirely on the positive, and that's the essence of positive deviance. Customers are invited to be on the lookout for times when hotel employees have gone the extra mile, or have done something special to make them feel more comfortable. It shows bright, smiling faces of happy employees, and it empowers customers to give rewards directly to those employees who deserve it most. This is also a way of setting customer expectations and creates a nice cycle of positivity (things gone right).

Figure 2.27 More examples of methods to increase customer value and revenue.

People are very skilled at finding fault, and it may be part of human nature to look for things out of place, or abnormal situations eliciting primordial survival instincts. We have been taught to find fault since our elementary education. How many teachers focused only on your strengths or unique gifts? Graded homework typically shows what you did incorrectly. By asking customers to be on constant lookout for times when employees do a great job, employees are also motivated and rewarded by the customers they serve.

Managers who once looked for negative exceptions are now put in the enviable position to be the conduit of positive reinforcement. Creating such a great reward mechanism demonstrates a profound understanding of human performance technology (HPT) by closely tying rewards to the most desired behaviors, which resulted in customer delight.

This hotel has also shown a great deal of integrity and it was very impressive to see in action. One trend many companies have is a "green" initiative (i.e., environmentally conscious). It's not only what you say that's impressive, it's what you

Figure 2.28   Hostess's name humanizes service.

Figure 2.29   Inviting open spaces.

Figure 2.30 Breakfast seating area.

Figure 2.31 Cows in the pictures make it clear that this is where the milk is.

Figure 2.32 Cold breakfast area with meaningful colors.

Figure 2.33 Visual methods for making it easier for customers to find eating utensils.

Figure 2.34 Standard signage demonstrates consistency and standardization.

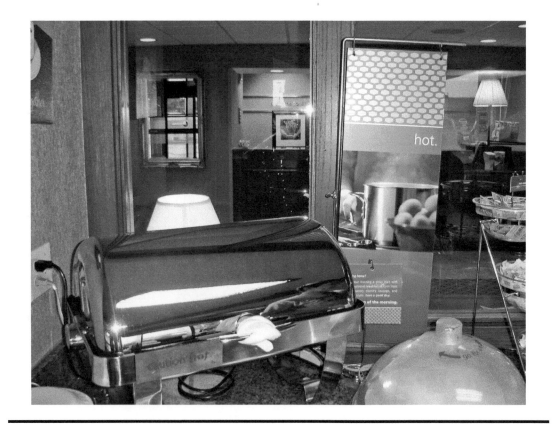

Figure 2.35 Hot breakfast with red signage.

Figure 2.36   Catch Me at My Best is awesome!

Figure 2.37 A biodegradable plastic-like plate is a first for me.

do that matters. This hotel actually applied green thinking and procured biodegradable cups, bowls, plates, and silverware (Figures 2.37 and 2.38).

Inspirational messages were prevalent, and these messages were always
accompanied by a picture. Images always have a deeper meaning, and you can
see just how engaging and inviting the messages are (Figures 2.39–2.41). It's more
about connecting with us on an emotional level to find what stirs our imagi-
nation. Using memories from childhood, we instantly shoot back to a time of
carefree days filled with play. As we ride the elevator, we remember the weight-
less feeling of swinging from a rope tied to an oak tree over a riverbank with
our friends in July. We remember the courage it took to let go at just the right
moment. Flying through the air weightlessly and plunging into shockingly cold
river water, we instantly get relief from the heat of a midsummer day. This happens in our brains while we ride the elevator, and by the time we arrive at our
destination, something changed; we now feel joy!

Now imagine it is the dead of winter, mostly dark, and very cold. There is
no better feeling than to wrap your hands around a hot cup of coffee or hot
chocolate when the house is still cold in the dark morning hours. Your hands
are numb from shoveling snow or ice fishing, and the coffee cup warms them
up (Figure 2.42). It's that image that invokes memories of the coldest day of the
year. It creates a personal connection with us through unforgettable memories.
A dichotomy exists between the heat of a dog day in summer contrasted with
the cool river water, or the freezing cold of the dead of winter with the comfort

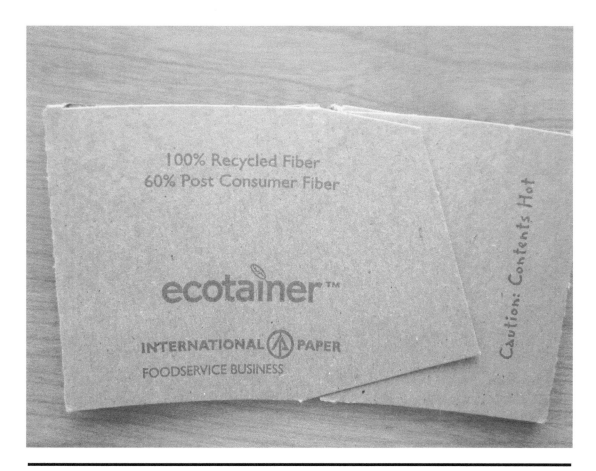

Figure 2.38 Reinforcing the value proposition with consistency and follow-through.

that a hot drink provides. The dichotomy is moving from a fleeting moment of discomfort to one of comfort. The value proposition is clear; we deliver comfort to the weary traveler. It would be impressive enough if only one location of the hotel chain had adopted these connections, but this has been observed in several locations.

While any single frame of our first Gemba walk might not be overly impressive by itself, when you take a step back and look at the entire Gemba walk we appreciate those efforts made to delight customers. A profound understanding of the customer condition is a recurring theme here. Of all the different experiences we've observed on our journey, what's most impressive is that someone has taken the time to walk a mile in the customer's shoes and actually do something about it! The people who designed this experience must have traveled themselves. Designers must have gone through the same annoyances at the airport as the customer experiences. Our hosts must also have waited in long lines and skipped lunch on the way to make a tight connection, and they've arrived in unfamiliar territory feeling tired, thirsty, and sticky. Obviously, they've gleaned tacit knowledge that can only come from direct observation. Having learned firsthand what it feels like to be a customer, a great deal of thought was put into designing an experience that makes it

Figure 2.39   Positive uplifting messages transmit a sense of helping one another.

Figure 2.40   Creating a sense of shared environmental responsibility.

Figure 2.41 **Inspirational message posted on elevator door ties to value proposition.**

Figure 2.42 **Invoking memories through imagery that resonates with guests.**

intuitive for someone whose condition may be disoriented, tired, and feeling jet lagged from travel.

Implementing a service design that makes it easier for travelers to get comfortable faster is impressive. So is creating an open space where you might even be comfortable enough to spend a moment with someone you've never met before.

Our host has obviously walked a mile in our shoes and listened to customers! That is why our first Gemba walk is named "Design for Customer Experience."

Let's reflect on how we use the Gemba walk to observe deeply from the customer's point of view. This versatile method, which had its humble origins in manufacturing, now has a good deal of promise for identifying new service delighters. In the case of the hotel, the service delighter had to do with designing the customer experience around the customer's condition, as we just learned. Imagine if you were running a hospital, which of the techniques that you observed in the first Gemba walk could be directly applied to the admission process?

I was working on the roof for my wife and for whatever reason I thought I was still young enough to just jump off the roof. I got down rather quicker, but I also dislocated my shoulder in the process. Try as I might, I couldn't get my shoulder socket to go back in correctly. Pain grew in intensity and I knew eventually I would have to go to the emergency room (ER).

The first thing I noticed when I arrived at our ER was a sign at the front door that said "After Hours Push This Button." No lights were illuminated, and it was Sunday, so I pushed the button and waited. Soon a nurse came to the door and opened it. Once she found out that I wasn't seriously injured, the nurse asked me why I rang the doorbell. I thought I was following instructions, but I sensed this had caused her some inconvenience. A great starting point would be to consider the customer's point of view, and a warm smile would go a long way to make a customer feel welcome.

Some people are actually terrified of healthcare institutions. The brightly lit clinical environment can invoke flashbacks from horror movies. No one wants to be told, "Go down the hall, turn left, take the elevator up, turn the right, and you can't miss it."

What happens during construction of a new wing? Construction can also make people disoriented in any environment. At the airport, we find ourselves suddenly walking down a dark narrow corridor with plywood walls, and signs have become sparse. Are we headed in the right direction? In the back of our minds we wonder, will we have to go back through security if we continue on this path? Will we become hopelessly lost never to find our way out of this labyrinth?

Notice the paper airplanes on the ceiling guiding us toward the exits (Figures 2.43–2.45). How many people might feel lost or disoriented without such a visual? Creativity engages us with a sense of awe, wonder, and enthusiasm of what fun things there are to do in Colorado in February.

Figure 2.43 Arriving at Denver Airport in Colorado at subway.

Figure 2.44 Creative and colorful paper airplanes point toward the escalator.

Figure 2.45   Guiding our way out of the subway and toward baggage claim.

Does our destination airport have a Smartphone application (app)? How could an app be designed for a particular airport? What messages could be communicated to arriving passengers? Smartphone GPS technology is awesome, so why not build an airport app with very cool features? Restaurant chains might also be interested in helping an airport design apps so the cost to the airport is minimal. This could easily make confusing airport maps a thing of the past. Who has a special for lunch today? You just arrived on a red-eye flight and you are famished. Where can you get breakfast now? Which line at security is fastest? See how much fun this is?

While we are dreaming, I want to be able to push a button on my Smartphone that will automatically find a decent hotel located in the direction we are driving, negotiate a fantastic deal based on inventory of unsold rooms, reserve it, and put in a waypoint on my GPS. Is this science fiction? We'll have to see.

Explore with your team (see Placemat 2.1): Let's take a moment to brainstorm a list of new potential delighters in any industry. Go for quantity, and do not judge the ideas. Once finished, review the lists to see if there are any new potential delighters that could be adopted in your business. Would any of these provide a potential *Wow*! from your customers?

One of the keys for any organization is to truly connect with customers, and it must clearly communicate its unique value proposition, which brings us to the next chapter. Let's explore how companies communicate, manage, and reinforce their value propositions.

Chapter 3

Value Propositions

The concept of value propositions ties into a company's brand; it's your unique pitch that will compel people to give your service a try in the first place. The experience your customers receive must be in harmony with the perceived value proposition. Exploring, updating, and refreshing your value proposition should be considered a mission-critical task for any organization. Companies have gone out of business when their value proposition no longer resonated with customers. For example, one video rental store's value proposition may have included a physical store. I remember trying to get a copy of the most recent releases, and standing in line with my wife and children on Friday nights. People really wanted *entertainment delivered*. An Internet startup's value proposition was more about the delivering process and eliminated the physical store. This proposition also eliminated a drive across town and customers waiting in line. The startup had more of an online presence, and it created a subscription-based model. While no company can avoid missteps, and at the time of this writing the startup was still in business, the rental store chain was in bankruptcy. Similar situations happened with bookstores. Once your value proposition no longer resonates with customers, it is only a matter of time before your organization evolves or dissolves. The world changes, yet many people still seem to think that what worked in the past decade will still work tomorrow.

What's your organization's unique value proposition? Once identified, how could it be communicated through the multiple channels with which your customers interact?

Value propositions need to be reinforced with consistent messages and actions before, during, and after the sale to demonstrate unity and integrity to your customer segment.

Integrity is not ambiguous. It really matters what is being promised to your customers. The minute customers have an idea of what your value proposition is, they'll be looking for signs that confirm (or deny) its presence. Perhaps the best way to illustrate value propositions is to examine a couple of companies. Let's go on another Gemba walk and share some observations to help explore this concept more deeply.

Take a look at the picture of the campground (Figure 3.1). Look at it carefully. What's their unique value proposition? The value proposition is actually right on the sign, and this begins to set customer expectations. Value propositions also define which market segment the Pine River Paddlesports Center (PRPSC) would

Figure 3.1   Value proposition.

like to serve. It also serves as a guidepost to prospective customers whose presence would likely detract from the enjoyment of others. Clearly some customers should be steered elsewhere. Perhaps loud fermented hops–drinking party animals would not fit into the vision of an ideal customer desiring a clean and quiet campground. And this sign puts those customers on notice even before they might consider staying.

My best friend and I were looking to stay at a campground located near a river where we could also enjoy canoeing. We really wanted to catch a few rainbow trout. Cooking fresh trout on an open fire connects us with nature in a deep and meaningful way. Furthermore, we wanted to stay at a place where we could have some privacy and enjoy a weekend out in the woods. As we pulled into PRPSC, the value proposition that was highlighted on the sign was reinforced as we walked toward the front office (Figures 3.2 and 3.3). Right on the front office door was a wonderful little sign that exclaimed "Fun Spot!" So even before we opened the door, we were already anticipating having fun.

Kayaks (Figures 3.4 and 3.5) suitable for navigating challenging river rapids were neatly organized and stacked in front of the office. The colors were distributed in a very engaging way. Neatness reinforces the value proposition of *clean*.

Imagine you are camping in a pristine national forest at a privately owned campground, and you step into the restroom for the first time (Figure 3.6).

Figure 3.2   Walking to the main office.

Figure 3.3   Fun spot!

Figure 3.4   Colorful and neatly displayed.

Figure 3.5   Organizing kayaks.

Figure 3.6   Campground restroom.

Imagine what you might expect when you open the door. Visions of cobwebs in the corners, screens rusted from moisture, damp and musty smells from the dark green mold growing on the shower stall, and dirt on the floor. Some child left their wet and muddy socks in the corner. Bar soap remnants left on the counter, which was imported from a basement far away. All these images are conjured up from our past experiences or imaginations.

What was fascinating about this visit was how clean the floors were in a campground restroom. My hiking boots collected quite a bit of sand from the campsite, and arriving at the restroom the floors gleamed. I felt badly about stepping on it with my hiking boots and actually wanted to take them off. Looking at the bottom of my boots, however, I realized my boots were clean and sparkling. There wasn't a grain of sand on them. Where was the sand?

I deduced that the irregular blocks of the decorative walkway had somehow flexed my boots sufficiently to knock the sand off (Figure 3.7). Was this planned? Talking with the owner of the campground, I found out he selected the texture of the walkway based on appearance. Removing debris was an unintended consequence or benefit of his choice.

It was surprising to see a live plant in the men's room. Sink tops were gleaming (Figure 3.8), the mirror was polished, and there were colorful signs inviting guests to play Frisbee golf or take a canoe trip (Figures 3.9 and 3.10). I was very impressed.

The thing that impressed me the most was the fact that we used this restroom over a period of a couple of days and we never saw a case when it needed to be

Figure 3.7   Walkway to the restroom.

Figure 3.8 Sink with a plant.

Figure 3.9 Colorful invitations and announcements.

Figure 3.10   Nice reminder with a thank you.

cleaned. It was always clean! This had me scratching my head because there was a rather large group of youthful campers visiting, and this was the only facility in the small campground. I never saw anyone cleaning, so how could it be possibly kept this clean?

At least part of the answer is preventive maintenance (Figure 3.11) of the campsites, where grasses are allowed time to recover; removing possible sites from use limits the number of campers at a given time.

The next day a young "cast member" took us on our fishing expedition. I asked him who was responsible for cleaning the restrooms. He said it's everyone's responsibility. Each employee who uses a facility is expected to clean it afterward. Wow! This demonstration of personal responsibility at a $20 a night campground was uncanny. I asked, "What happens if someone slacks?" The employee replied, "The owner's wife checks up on us, and if somebody isn't pulling their own weight, they're going to hear about it." That led me to wonder how they go about recruiting talent at a campground, and I asked the owner (Mark). Mark said they're very picky about who they hire. I asked Mark what he does when a cast member just doesn't cooperate. He just said, "We find a replacement."

The big delighter at PRPSC campground was how much integrity the staff showed by aligning to the value proposition with a sense of mutual accountability.

Figure 3.11   Preventive maintenance (allowing grasses to regrow).

Let's take a minute and reflect on the second Gemba walk; what ideas, concepts, and thoughts might you have that could be applied to your own business? Let's start with the first question: What is your company's value proposition? How is your company's value proposition communicated? How well is your company demonstrating congruency to your stated value proposition?

While visiting Ann Arbor, Michigan, during the summer, my car happened to be due for an oil change. I asked where I should get my oil changed since I would soon be returning home, which was about a 700-mile drive. My best friend's wife told me she used Uncle Ed's Oil Shoppe (Figure 3.12). I cruised into the oil shop parking lot and could easily see by their sign that the value proposition was "fast, friendly, and professional." Even the name *Uncle Ed's* implies trust and experience. Who wouldn't trust their own uncle to take good care of their car?

Fast implies that they will value my time, *friendly* implies that employees treat customers with respect, and *professional* implies a high degree of competence. I would expect the work to be done right the first time and this performance characteristic values my time since it should not require a return trip.

The place was busy and there were multiple service bays where customers' vehicles were already parked. Immediately upon pulling into the parking lot, a young man emerged and made eye contact with me, and then gestured where I should wait. After I stopped, he came up and greeted me. He said he would return in a minute and disappeared into the shop. Less than a minute later

Figure 3.12   Uncle Ed's Oil Shoppe value proposition creates expectations.

he returned as promised. He introduced himself and asked, "What services can we provide you today?" I informed him I just wanted an oil change, and he directed me into one of the service bays. He obtained basic information about my vehicle specifications, and introduced me to my service advisor (Robert).

Robert was a big fellow, but he was also friendly. He smiled and started talking about my car. A Lancer Evolution is a family sedan equipped with a manual transmission, carbon-fiber wing, a rather large intercooler in the front, and has all-wheel drive. It looks like something many young adults would want to drive on a night out. I purchased that car when it was new, and always took good care of it. Suddenly I was surrounded by people who shared the same passions about cars, and that common denominator opened the door to some interesting conversations.

You don't have to get out of your car at this particular shop. Hence, we avoid the dreaded waiting room that some competitors have. Oil change waiting rooms typically have filthy carpet, worn-out magazines, and a neglected coffee pot that's literally on fire. The picture wouldn't be complete without a couple of crying children and sticky furniture.

By sitting in my car, I could listen to music or talk on the phone. The employees seemed to be working in unison, with each person having defined roles, and each person supporting the next. As soon as one worker completed a specific task, they would spend their time talking to customers and building rapport. No one was engaging in internal coworker chit chat like you might observe at some retail shops. As an extreme contrast, at one hardware store I actually observed employees complaining about customers over their internal radio system.

It was amazing how well I was kept informed at this shop. As soon as Robert had a job order confirmed, his enthusiasm for making the sale was obvious and he used his booming voice to announce the sale so that everyone in the facility could hear. Besides adding excitement, his announcement was very specific in that it informed his peers exactly which service was just sold and in which service bay. Neighboring customers could hear what I just purchased and might be encouraged to do the same.

Since a customer was sitting in their car while a technician was working underneath the hood, there was a barrier between the customer and the worker that prevented the customer from seeing what was going on. A customer could hear somebody wrenching around under their hood, and might be concerned about unusual noises. In fact, drivers are attuned to unusual noises coming from their vehicles. To address any potential concerns, the technician announced every operation as it was occurring. This kept the customer and crew working beneath them in the lower level (underground) well informed. It was amazing how much exceptional communication was occurring between team members and the customer. By doing this, customers are kept apprised of the status during the entire service process, which also helps keep their mind off of the clock.

This had the effect of turning a boring operation of waiting and not knowing the status into a fast-paced pit crew–like experience. When promising *fast* service, it is more important to realize that a customer's perception of time is different than actual time elapsed. For a bit more context about *perception of time*, visit a Disney resort and see how they have made waiting for the ride part of the overall experience. Even signs tell customers how long the wait will be, which helps to set expectations. My daughter informed me that she noticed that actual wait times were consistently less than the estimates.

Distraction also seems to work for administering a vaccination; the nurse pinches your arm to confound your nerves, which minimizes the feeling of sharp pain from the actual shot. Managing customer perceptions by keeping their attention focused elsewhere is effective and fairly inexpensive.

Robert returned to show me the recommendations. It was amazing that within a very short amount of time their computer system had generated a very professional-looking artifact for me to review. Robert also presented the recommended services in a fascinating way. He got down on one knee and presented it like a waiter would present you a fine wine list at a five-star restaurant. What he was doing was equalizing the power differential between us by putting his head lower than mine. Robert wasn't towering above me badgering me into buying services I didn't want to buy. Instead, he was presenting a menu of options to me in a professional way to give me the power to choose. It clearly put me in the driver's seat, and I was driving. He was only presenting relevant options to me in a very professional way, and was describing how each of the options would benefit me.

The computer had selected a couple of add-on services which I wasn't quite sure about. Sensing my reluctance, Robert asked me to think about it for a minute while he checked on another customer, and took the immediate pressure off of me. By walking away, he continued to show respect, and he demonstrated that he could "read" a customer and opened a space for me to consider additional services. By the time Robert returned I was much more receptive to his suggestions. One suggestion was the fuel injection service, which he said would increase fuel mileage while reducing emissions. Robert described how they would hang one liquid inside my hood that would enter the throttle body area of my intake. Another chemical would be added to the gas tank and when the two liquids met, look out! Pow! He hit his hand with his big fist to accentuate his point. I was intrigued. Robert was empowered to offer a discount without additional approvals, so I was up-sold something I would normally never buy. So I decided to give it a try.

Add-ons or up-sales have a huge benefit to any service provider. Up-selling increases revenue without adding the administration cost of admittance, thereby increasing profits. For example, when replacing an engine water pump, it takes only a little extra time to swap out the thermostat and replace the hoses. Prices are generally fixed on what it costs to remove all of the parts in the way of the water pump, and in my experience, up-sale savings were rarely passed on to me. Once a vehicle is in the service bay, the check-in process is no longer needed. A customer is already getting some work done, and by telling that person it will only take a little additional time, they might be receptive to an up-sale. Receptivity increases if customers are reasonably confident the service is going to be done right the first time, and if the people around them are friendly, outgoing, and take time to build a relationship with the customer.

While my car was being serviced, I continued to be impressed by how professional the team members were. They truly knew what they were doing. People working in the lower levels would come up and talk to a customer when they weren't busy doing something else. This was an example of living up to their promise of friendliness. I kept wondering what drives this team of people; what is the underlying motivation behind it all?

At the time, my car was idling and one of the technicians approached me and told me that there was a chance that a *check engine light* would illuminate while they were cleaning my fuel injectors. He told me this was normal and if it happens, it's nothing to be alarmed about. He went on to explain that the chemicals they were putting into the throttle body might be detected by the oxygen sensor and cause a fault message. He told me that he would clear any faults that happen to the car's computer.

My car started running a little rougher and it was at that time I looked in the rearview mirror and clouds of smoke were coming from the tailpipe. I could smell sulfur compounds coming out of my throttle body. All of the accumulated junk from gasoline additives was being burned off by my engine. Robert came over with a big grin. He reminded me about his description, and we both chuckled for a while. I was almost embarrassed with how smelly my exhaust had become and noticed that some of the customers in the adjacent bays were closing their windows—too funny.

On my way out, the artifact I was given had an online survey that enticed me to provide feedback about my experience by offering an incentive for my next visit. Everyone in the shop took time to thank me and said goodbye. They all knew my name!

As I left, I was regretting I had turned down one of the optional services. This service provider had done such a great job with treating me the way I wanted to be treated and living up to their promised value proposition that I couldn't wait to return and spend more money there! If I had enough time I would've done a U-turn and gone back.

Service excellence is all about making customers feel good about spending their hard-earned money and wanting to come back. Humans are social creatures and our need for belonging is great. We want to be accepted by others. We want to be respected above all else, and we want to trust the people we're paying. Uncle Ed's Oil Shoppe had just redefined what a customer service experience should be.

Explore with your team (see Placemat 3.1): So how do you communicate your value proposition? Once you have made a sale, how do you reinforce your value proposition? After the sale, how do your employees communicate and demonstrate value and integrity?

What were some of the service delighters from the oil change facility? Which new service delighters might you build to apply to your own business?

One of the challenges that I would like each person reading this book to accept is to try this yourself as you go through different services. Take a Gemba walk and take a mental picture of your own experience, and look for positive deviation. What were some of the things that delighted you as a customer? Which service characteristics are transferable from one service industry to another without a huge cost?

My fascination with service has led me to want to thoroughly understand the essence behind service excellence, so I've identified five main properties or essentials for service excellence, and that's the topic of our next chapter.

Chapter 4

Essentials for Service Excellence

Service is filling someone's need—it's a human transaction. It's a pact between the helper and the recipient (customer). In that pact there is an expectation to provide a sufficient value or experience, which leads to a mutually beneficial relationship. The short-term goal of any service provider should be to make people feel good about spending their time and money. Another goal is to earn repeated business, but the long-term goal should be customer advocacy leading to word-of-mouth advertising. It is extremely powerful to have someone suggest your services to their friends and family. Creating the conditions that increase receptivity for an up-sale is also important since it costs less than a cold sale.

In the healthcare business, it is difficult for me to understand why patient preventive medical screening options are offered infrequently. For example, when a patient arrives and needs x-rays to verify if a bone is fractured, could some preventive medicine be offered like a flu vaccination while waiting in triage or sometime before discharge? There may be good reasons that healthcare does not often offer up-sales, and insurance companies may have influence over what is offered. Everyone wants great treatment at a low cost, but one could argue that the cost of an infection could be much more than the cost of the vaccination itself.

At the heart of service is a helping relationship and like spokes on a wheel (see Figure 4.1). Managing customer expectations requires an understanding of the elements of human performance, creating and living your team roles, and finally organizing for excellence. These factors seem to be the key elements for providing service excellence.

There's a big difference between *listening to respond* and *listening to understand*. To illustrate this distinction, think of a time when you attended a meeting where no one seemed to listen to anyone. People may seem polite on the surface, but are privately thinking, "If only that jerk would finish, I could make my point." As soon as one person is finished, another participant makes their point.

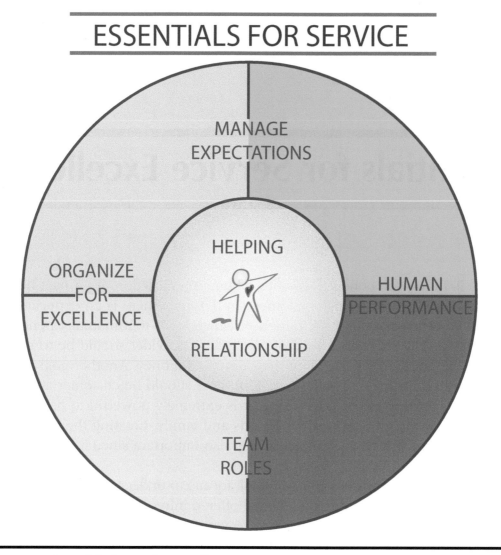

Figure 4.1 Essentials for service excellence.

Dysfunctional behavior repeats until they run out of time. Besides feeling relieved that the meeting ended, I'll bet you would be hard-pressed to recall the business outcome of that meeting.

As you pull information from your customer, you must use all the techniques available to you for listening. While flying back from Japan on JAL airline, I spoke with a flight attendant. What was remarkable was how she made me feel during this encounter. She moved close and looked rather deeply into my eyes during conversation. She was evaluating my reaction to her questions and by watching microexpressions on my face as I spoke. Her eyes were darting around as she carefully watched my eyes and face for nonverbal clues. I could tell she was skilled in understanding nonverbal expression, and with a language barrier it was even more important. I felt truly respected! Listening is a skill that is normally not taught in schools, and it takes practice.

We talked about respecting people. We talked about listening. We talked about observing body language, and we talked about pulling from your customer's needs. When your organization takes an order from a customer, how well does the face of your company do? Can a Web site or Web page be the face of the company? How skilled are your employees at recognizing the difference between pushing and pulling? How skilled are service providers at reading nonverbal communications? How could you find out for sure? Most companies take customer surveys, but how many take customer Gemba walks?

At the heart of service is the helping relationship (see Figure 4.2), and this is because helping one another builds trust. I have to trust you before I'm going to allow you to help me. To illustrate this point, think of a time you were in a vulnerable position where you had to rely on someone else for your daily needs.

HELPING RELATIONSHIP

"HOW CAN WE SERVE YOU TODAY?"

● EQUALIZE POWER DIFFERENCE

● ACTIVE LISTENING

FROM: PUSH   TO: PULL

 BUILDS TRUST

● OPENS COMMUNICATION

Figure 4.2 Helping relationship. (From Edgar Schein. *Helping.* **San Francisco, CA: Berrett-Koehler, 2009. With permission.)**

Think of a time when you were totally incapacitated with an illness or some injury that landed you in the hospital. If you have been fortunate enough to miss out on those experiences, think back to your earliest memories of your own childhood. These were the times when we were at our most vulnerable. We had to trust and rely on others to take care of our needs.

By putting yourself in a vulnerable position, you can gain a deeper perspective of humility. Here in the United States, there was a high level of teen pregnancy. For a time, high schools required female students to care for a surprisingly realistic infant doll. The doll had certain needs, and these needs were communicated to the student at random intervals. The doll recorded how well the student cared for it. It required attention (feeding, change of clothes, sleep) even during classes. The point of this was to help students make informed decisions on behaviors that could lead to a birth. Knowing ahead of time what the long-term responsibilities are can help anyone make a better choice, and infants are particularly vulnerable.

Once my family visited the Caribbean and stayed at a resort in an otherwise Third World country. My 10-year-old daughter was in heaven. She was in her own world, splashing in the pool, playing in the lazy river, and having the time of her life. It was wonderful being on an all-inclusive vacation; parents could lie on the beach away from their children, and not even be concerned about their safety.

That all changed when a resort employee hurriedly approached. With wide eyes, she informed us that our daughter had turned very ill, and we needed to rush her to the infirmary. When we arrived at the nurse's station, my previously healthy daughter was stiff as a board. Her lips and skin were pale white. She looked like the life was draining out of her body, and there was nothing I could do. She always looked healthy, and it was a huge shock to see her so ill. I asked about her symptoms, and was also told that an on-call physician was on his way over, which gave us some comfort. The doctor also prescribed medication to fight an infection, and sent her back with us to the hotel room for rest.

However, the tropical climate was hotter than we were accustomed to, and my daughter started showing signs of dehydration. Her condition was not improving, so I was getting very concerned about her health and wondered about the quality of care she had received earlier that day.

She kept getting sicker, and I was advised to take her to a modern medical facility in town. I carried my daughter downstairs to the lobby, and hired a taxi to take us to the facility. No one would accept a credit card or an insurance card, the only thing they would accept was hard cash. The doctor at the medical facility was very professional. I asked him to give her fluids intravenously, and he put me in my place and said that my daughter would be given a choice for her treatment. We were both totally dependent on someone else, and that's when we realized how interrelated everything is. In two days time, my daughter was completely back to normal, and I am very grateful that we found decent care for her.

Within any organization, if people don't trust each other they won't communicate openly. People won't share everything on their minds because they are not sure how that information will be used, or if the information will be used against them. Why would anyone risk telling you what's really going on if they don't have trust? The same thing goes within any customer relationship. When I take my car to a service station, I'm relying on those experts to tell me what's wrong with it, and to not take advantage of me. If I feel like I'm being taken advantage of because of my ignorance of the mechanics of an automobile, I'm certainly going to be less likely to take it back to the same facility.

However, if I'm treated with friendly conversation and shown respect, then I'm more receptive to an up-sale, thus more likely to be a repeat customer. Additionally, if the experience is positive enough, a customer may become a conduit to spread a positive experience. This is like planting the seeds for a good reputation. I want to believe that you have my best interests in mind.

The helping relationship corresponds to the heart of our bodies. It's our most human moment, and it's a time of vulnerability where we have to take a leap of faith. We all want to trust that someone won't post raw, uncensored full body scan images from airport security systems, so we continue to allow ourselves to be vulnerable in order to be kept safe.

Explore with your team (see Placemat 4.1): How you communicate your value proposition is central to meeting the expectations of your customers and providing service delighters. You've heard of "underpromise and overdeliver" when trying to keep in the good graces of your boss. Managing expectations is just as important with your customers as it is with your boss.

The same principle applies to service. At the oil change facility mentioned earlier, Robert asked, "What do think of that?" as my car billowed smoke throughout the facility. He made it clear that was the *pow* he was talking about. Workers kept telling me the status of everything. When a service was added to my order, Robert announced it clearly for everyone to hear. I was informed how much longer it would take, which ties back to the value proposition of fast, friendly, and professional. Frequent updates are a key to managing expectations and always trying to pleasantly surprise your customer.

I recently bought a pleasure boat. It did not run like I expected it to, and I found a local respected outboard mechanic, Ben. Once Ben had a chance to diagnose the motor, he informed me that it might be expensive to repair. He advised me to do the minimum repair to start with. When he called I expected the worst, but he thought he could repair it. He kept in touch and took the time to explain what he found. I followed his advice, and he was able to salvage the engine for 4 percent of the cost of a new one. When I got my boat back, the engine looked and ran like new. He had milled the heads, which added compression and horsepower, and painted the block. By cleverly managing my expectations I felt grateful to have such a professional service. Had Ben estimated a small repair cost, and then had to exceed it, it would have been a source of dissatisfaction. He mastered the art of managing customer expectations (see Figure 4.3).

MANAGE EXPECTATIONS

● **COMMUNICATING YOUR VALUE PROPOSITION**

PROMISED VALUE

DELIVERED VALUE

● **KEEPING CUSTOMER INFORMED**

● **PLEASANTLY SURPIRSE**

Figure 4.3   Manage expectations.

What examples do you have when someone pleasantly surprised you? Now would be a great time to brainstorm a list of possible service delighters.

To help your team get started, here is an example. In a world that seems to have adopted e-mail as a primary source of communication, when I get a handwritten note from somebody thanking me for buying a service, it makes a real impression. Could a timely handwritten note showing sincere appreciation be a source of delight for your customers?

How could you do a better job managing expectations and communicating with your customers? When I think of healthcare, I think of a series of disjointed processes. It could be that the design of the healthcare experience is built around very specialized expert caregivers instead of around the patient.

You may have had different experiences, but it might not hurt to look at things from a customer's perspective. Is it really necessary to move the customer all over the facility just to wait for 20 minutes at each stop? Is it really necessary to ask the customer if they smoked cigarettes multiple times during one trip to the hospital?

Explore with your team (see Placemat 4.2): At the heart of any high-performing organization is a deeper and profound understanding of what makes human performance tick (see Figure 4.4). Engaging an individual's head, heart, and hands is a key to creating a culture of high performance. Inspiring people to perform is central to engagement, and it all starts by creating a vision that resonates with

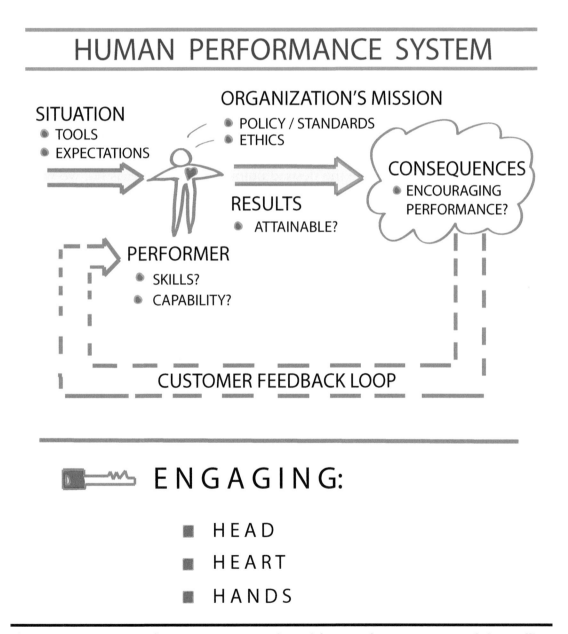

Figure 4.4 Human performance system. (Adapted from Andrew Longman and Jim Mullins. *The Rational Project Manager.* **Hoboken, NJ: Wiley, 2005. With permission.)**

people. Ask yourself why you joined the organization you joined. Why are you in the industry you're in? What was it that attracted you to this field in the first place? Is there any difference between what attracted you to your industry and what the current reality is? If so, what are the impediments in the way of you fully engaging?

As we look into human performance, people tend to focus on training first. "Things would be better around here if we only trained more." "Let's retrain everyone; it has to be what we need to do." One of my mentors explained to me that training is always a safe bet. It can hide weaknesses in other areas of leadership. It ignores the interrelationships that people have among different departments. It's easier to sweep things under the carpet than diagnose and address the real root cause for a performance problem.

Sure, training is important, but it's not the be-all, end-all to address human performance. I think there's something much more important, and one element of human performance is consequences. Are consequences aligned with the work standards, company policy, and vision of the organization?

Comparing the output that a performer produces with the work standards should result in consequences. Consequences need to reinforce desired behaviors, and as needed discourage undesirable effects. The first question is can "average" people be expected to achieve the desired output? If the answer is no, we will want to find out why. Many times we find that people don't have all the tools they need to do the job customers demand.

My oil change experience suggested that the service adviser was supported by rock-solid systems and procedures to support service excellence. This was evident by the professionally suggested recommended maintenance work that was custom tailored to my specific vehicle. The color printout artifacts I was presented made the suggestions even more compelling. They also had all of the needed parts on hand to perform the services, even though my car was a somewhat rare model.

If you find that you don't have the process capability to support flawless execution of the frontline staff, this is a great first place to start. What are the consequences if something goes wrong? How will somebody know if desired output was not achieved? This brings us to customer feedback. How does the performer get customer feedback?

What kind of feedback do they get? How often is performance (or customer) feedback communicated? Recall the concept of *unconsciously ignorant*. The last thing anyone wants is people thinking they are competent when they are not. Someone needs to have the guts to tell them the truth.

Most often performance feedback comes from the boss, and it has his or her perspective built in. Unfortunately, feedback is considered inherently negative and is not always the best way to achieve the goal of increased awareness and performance. *Feed-forward* is a preferred method for improving capability. For additional information on feed-forward, please refer to Marshall Goldsmith's book *What Got You Here Won't Get You There* (2007, Hyperion).

Does the performer have the skills and capability to do the work and all the tools needed? Do they have all the information needed? Do performers have the skills? When we look at human performance from a holistic standpoint, it begins to make even more sense. The key point in this is to take a deep dive and look at what could be going on with the system of work that could lead people to nonperformance.

Explore with your team (see Placemat 4.3): During the oil change, people were always changing roles (see Figure 4.5). For example, the person whose job it was to remove the oil filter underneath a vehicle was not busy all the time. There are times in any service delivery when people have some time available, and it's during these moments that service providers could open an opportunity to build relationships with their customers. A helping relationship is one key to building trust.

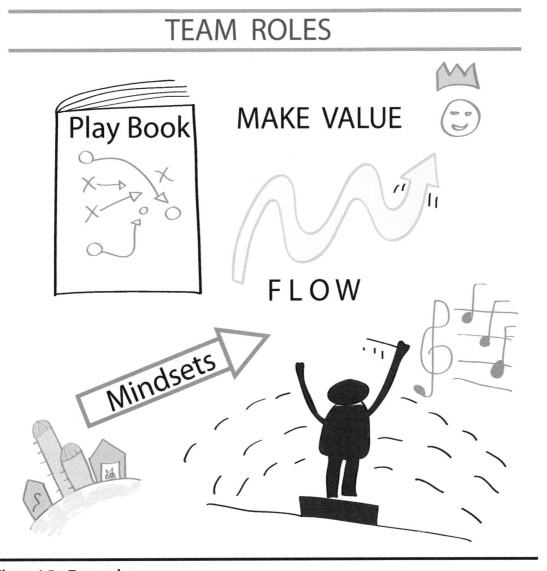

Figure 4.5   Team roles.

What motivates anyone to climb out of the pit below my car and strike up a conversation with the customer? There has to be some kind of playbook being used to coordinate workers' activities. Team members are encouraged to build a rapport with customers, and this rapport softened customers for an up-sell. If an oil change facility can master this skill, what about your own organization?

There was a time when functional experts were employed by organizations. Functional experts worked in silos and the mindsets were difficult to penetrate. We are really all in this together. It begins to make sense that team roles are all about orchestrating excellence. It's about having a decent playbook. People know what they're supposed to do and when.

What are the most important things in developing service excellence? Prioritizing these actions based on a playbook is a great step. The ultimate goal is to make value for the customer. If your value proposition is *fast*, then your ideal customer is someone who is very busy; a customer that values their time perhaps a little more than squeezing every dime. A customer who would prefer to change their oil themselves is not an ideal customer. Nor is someone who would prefer to wait a very long time at a discount shop an ideal customer. They just don't fit the target customer—people who value their time more than money, and are willing to pay a little more for convenience.

Explore with your team (see Placemat 4.4): Organization charts have been around since the train wrecks in the 1800s to hold someone accountable. This caused the blame game and the organizational chart focuses our attention in the wrong place. It focuses us on thinking about the boss and what's good for the boss. Will this make the boss look good, or will this make us look good in front of the boss? These questions cause all types of dysfunctional behaviors; backstabbing colleagues, a scarcity mentality, and a focus on the top. What's missing from that picture is the customer. The inverted triangle (see Figure 4.6) has been around a while, and it brings up a great point. Management's job is not about telling people what to do; instead, it's about removing the roadblocks that get in the way of people doing the work on a daily basis. Senior leaders are now starting to get the idea of their new role, and a big part of that new role is that employees are the boss's first customer!

Organizations that focus on the traditional boss mindset missed the point. The question leaders need to ask is, "How good of a job am I doing supporting and enabling my people?"

How do you show appreciation of your employees, peers, supervisor, or customers? How does leadership appreciate you? What are the different ways in which you're genuinely appreciated?

Lean leaders lead by asking questions. They treat people with respect and show appreciation for bringing a unique perspective. Not knowing the answer is a sign of humility, and inquiring must be done in an authentic way. Pulling all these different characteristics together, it might be called servant leadership.

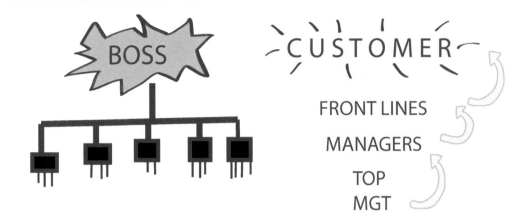

ORGANIZE for EXCELLENCE

BOSS

CUSTOMER

FRONT LINES

MANAGERS

TOP
MGT

BEHAVIORS:

LOOKING UP

INTERNAL COMPETITION

SUB-OPTIMIZING

BUREAUCRACY

FOCUS ON BOSS

BEHAVIORS:

LOOKING ACROSS

COLLABORATION

ENABLING

EMPOWERING

 FOCUS ON CUSTOMER

Figure 4.6   Organize for excellence.

Explore with your team (see Placemat 4.5): Next we are going to explore the mindsets needed for service innovation to flourish. Gemba walks are used to find a steady stream of new ideas on the path to create customer delight through service excellence. A practical way to sell new ideas is provided with innovation plans, and a method for communicating with the rest of the organization is outlined.

Chapter 5

Service Innovation Mindsets

Creating the conditions for service innovation to flourish should not be taken lightly. In fact, creating a proper environment is one of the biggest challenges you may undertake. Allowing people to experiment with the ingredients in a service recipe that took long to perfect is always difficult. The way I approach this is to let people know that *anything can be improved*. In fact, I will create a large poster, and put that headline in a conference room for a team meeting and let everyone know that it's expected that we start challenging the status quo.

Another force that innovators will have to reckon with is organizational inertia. If things are going well, there will be a natural resistance to any change. You've heard the old saying, "If it ain't broke, don't fix it," and that message rings as clear as a bell. If things are going relatively well, why would anyone take a chance with possibly degrading our service? It's true that you could make things worse—but that's conventional wisdom talking! Changing mindsets takes practice, and you are not going to break out of any mental barrier by just thinking about it. You have to act your way into a new way of thinking, and not the reverse. We have to open our mindsets to the fact that the world has changed, and it will never be the same. We all learn by doing!

Companies at the leading edge of innovation have at the core of their strategies a desire to change quickly. People understand the importance of supporting a culture of rapid experimentation, and their rewards and consequences support this core belief. If your company is small, you have an advantage over large companies when it comes to experimentation. You don't have to deal with layers of entrenched bureaucracy. If you happen to work for a large company, you definitely have your work cut out for you. If innovation and experimentation are not widely recognized as the means to create a competitive advantage, you will have to help your leaders connect the dots first. Most leaders are held accountable for certain performance metrics, and it will help to be able to tie these prerequisites to leader goals. In this way, leadership will recognize that service innovation is one of the keys to achieving their bonus. When that happens, you may find a very receptive audience with these ideas of experimenting with the service design to create customer delight.

Empowering within boundaries might help open the door to allow innovation to occur. Perhaps instead of getting free range for everyone to go out and make changes to the service delivery without any approvals, it might help to provide a guideline or framework for people to innovate within. The trick is to balance the speed of an environment steeped in the spirit of creativity and agility with breaking the effect of an approval structure. The sweet spot just might be using a simple one-page innovation plan.

The innovation plan (see Figure 5.1) is intended to help people to propose changes while obtaining input and approval from key stakeholders. It is also used to communicate how these will be evaluated to measure the impact on customer experience. Innovation plans are carried around and used to communicate with stakeholders on a personal basis to obtain needed input. Then they are posted on a team wall. Quick stand-up team meetings can be held to discuss the status of each innovation plan and to measure and monitor performance. Imagine the joy that completing a weekly team meeting in twenty minutes would bring to your staff?

Never punish! You've probably heard the old saying, "once bitten, twice shy." As soon as someone gets punished for experimenting with the service process, your effort will be pretty much dead in the water. Tying this back to the human performance system, we need to make sure the consequences and rewards support an environment that encourages experimentation and some risk-taking. We also need to empower people to regularly take time for improving the work. I normally

Innovation Plan	Owner, Date, Approval
Potential new delighter(s) • Describe what is being suggested. • Describe where this idea was found. • What suggests this is a possible delighter?	**Test Plan** • How will this be communicated? • What job aids are needed? • What level of support is needed? • How will customer feedback be obtained? • How will we communicate results? • What is the contingency plan? • What are the costs vs. benefits?
Impact on Ultimate Customer • What kind of feedback do we expect? • How will this provide a competetive advantage? • How will you know if this had the desired effect?	
Impact on Our People • What changes are needed? • What tradeoffs are there? • What objections are anticipated? • What level of support is anticipated? • How will training plans be affected?	**Follow-up** • Who needs to sign off? • When will this occur (Timeline)? • How often should progress be measured? • If successful, how can it be replicated? • How will this be standardized?

Figure 5.1 Innovation plan.

communicate this concept by hanging another poster in the conference room, and this one is titled "Improving the Work Is as Important as Doing the Work."

Before innovating for new delighters, you will need to look closely at the current level of customer satisfaction from your current service process. If your service is delivered inconsistently or if there are many complaining customers, then addressing this should be a top priority. By addressing the fundamentals of your customer experience first, the journey is started on the right foot.

Seeking new service delighters is an important part of improving customer experience. Therefore, freeing up time to allow for Gemba walks is going to be an integral part of making improvements to the service design. Management must believe this concept! It doesn't matter what is said, it only matters what is done, and therefore management must walk the talk when it comes to enabling people to be successful.

Gemba walks with your team can open the door for building a culture conducive to innovation. For one thing, you can have a lot of fun just selecting which company or service provider to visit. Even choosing which team members to invite on the Gemba walk sends a clear message and encourages people to reach for excellence.

Taking a few hours out of three or four team members' schedule demonstrates a real commitment to moving the needle. No matter who goes on the Gemba walk, it's important to share the key observations and have a quick debrief with your home team. By keeping everyone informed and involved, you invite commitment to making this work. How frequently to take Gemba walks in your organization depends on your team's ability to absorb new concepts.

Taking a Gemba walk in the service industry is unique, and you can learn much by just listening to people share their impressions from recent service experiences. In fact, you might not always have to take a trip to gather useful information. To illustrate this point, here are five very brief examples.

Example 1: Valuing Customers

A friend of mine shared her recent experience with a grocery store. This store uses a *frequent customer card* system that offers in-store discounts in compensation for identifying her as a unique customer and logging her purchases. After returning home, she reviewed her receipt and found a discrepancy. Upon calling the store, she was not very satisfied, so she decided to visit. As soon as the manager scanned her card, he suddenly became very helpful. We both concluded that the store manager recognized how valuable she was. My friend told me that now when she shops in that grocery store everyone welcomes her like she is their best friend.

How could your company leverage information better to assist frontline employees appreciate the value your customers bring?

Example 2: Acting on Customer Surveys (as in Right Now)

Can customer surveys be turned into a delighter? Definitely! A friend of mine works in a company that offers a small incentive to customers for filling out a survey after a visit. Her company decided to act on any customer survey that was below 80 percent satisfaction. She would call each customer to address causes for lower than desired feedback scores. Customers were very surprised that anyone would even read the surveys let alone call them, and this became an instant source of delight! As a result, she gained a steady stream of helpful hints, and by sharing these broadly within her organization, she helped accelerate customer service satisfaction.

In fact, she reports that they now have a system in place that a district manager gets a notification on his or her Smartphone the second any customer submits a survey. A detailed receipt is attached to the survey so that the manager knows who the customer is, what services were delivered, at which location, and most importantly, how to contact the customer. All this happens so fast that in most cases a customer is called within five minutes of submitting a survey.

As a result of a relentless customer focus, my friend began to question the logic of an 80 percent satisfaction score goal. She convinced others to call all customers who did not give a 100 percent score to find out what more they could do. My friend tells me that each day they receive 50–60 customer feedback surveys, and nearly all of them have glowing comments about how great a job her company is doing. In her words, "Customers tell our story so much better than we ever could." These comments are proudly posted on their Web site, which makes for some very authentic marketing.

What is your current customer satisfaction goal? How quickly do you respond to a customer survey or contact? What would be the benefit of calling a customer with a 100 percent satisfaction score (just to thank them)?

Example 3: Practicing Positivity at Work

Having a nice day? Visiting the right service provider during your lunch hour or after work can actually brighten your whole day! One company implemented a "credentialing" process whereby each team member gets to know their colleagues' professional strengths. They practice introducing each other by acknowledging the high points of the other person's credentials. This creates a positive team environment where people say good things about each other, even when the person is not present. It takes a lot of practice because people are so accustomed to being competitive and tend to focus on weaknesses.

My friend tells me that during training sessions she will stop someone in their tracks if they start saying anything negative. She asks them to be silent until they can say something positive about another person. Eventually people become proficient at being positive and appreciative of the knowledge, skills, and abilities their coworkers bring. When a customer enters such a positive work environment, it contrasts with their normal workplace so drastically that they want to hang around, just to be in that atmosphere. Customers appreciate positivity by posting comments like, "Your company just made my entire day!"

What could be the benefit of creating a more positive service environment for your team and customers?

Example 4: Valuing Time

Earlier in this book we introduced the idea that customer expectations change over time and this causes new delighters to erode into performance or basic characteristics. Service companies understand that time is of huge value to customers; however, it seems this requires an ongoing concerted effort. Companies who can help reduce waiting time will have a competitive advantage over those who do not. We illustrated this point with the pager coaster (see Figure 1.3). A variation on this theme is to prevent customers from having to wait by allowing them to check in ahead of time. Even airlines have gotten the message and now allow customers to check in and obtain a boarding pass ahead of time (see Figure 5.2). You can even get a two-dimensional bar code image e-mailed to your Smartphone that eliminates the need to print a boarding pass. Now even barber shops are getting on board with saving customer's time by encouraging a virtual check-in.

In what new or innovative ways could you save time for your customers? How does saving time tie into your value proposition?

Example 5: Enabling Changes That Tie into the Human Performance System

At one time, airports charged customers for using short-term parking to drop off passengers. At the same time, officers of the law were spending time and energy keeping traffic moving, and discouraged people from loitering at the drop-off point outside of a check-in area. In fact, citations were swiftly given out to anyone who dared disobey any parking ordinance.

Budget-conscious people who wanted to avoid parking costs created additional congestion near drop-off areas. Enter an enabling change: short-term parking is now free for up to 30 minutes! This one change stopped penalizing people who wanted to avoid the congestion of a crowded drop-off area. It also allowed time for families to properly send loved ones off on their way. Hence, a new delighter is born! Enabling changes by tweaking policies and consequences can help to align human behaviors toward a more desirable state.

What kinds of enabling changes might you consider to address customer (or employee) pain points?

Figure 5.2 Online check-in example.

Innovation Plans

There never seems to be enough good communication in organizations, and now we are talking about using Gemba walks to identify new service delighters and then experiment with implementing changes. One solution to the communication concern is to begin using a standard format for seeking input and approval for the new ideas that you and your team are going to suggest. The innovation plan is a great way to sell your ideas in one page and is derived from the Lean tool called *A3*. The constraint of having to boil your ideas down to fit in one page forces clarity. It's something that people can readily digest, and by posting innovation plans in a conspicuous area, people can interact with the idea of change at their own pace.

The innovation plan is divided into two sections. The leftmost column communicates what the idea is and the purpose behind it. This format allows other people to provide their own input, and helps you to communicate what experiments are planned. The right side of the innovation plan has more to do with the timing for trying the ideas out, and of course, testing, monitoring, and adjusting.

Leading and Lagging Indicators

As stated in earlier chapters, new service delighters erode over time and eventually become a customer expectation. Once your organization becomes proficient at implementing changes and becomes supportive of people experimenting with the service process, then you are well on your journey. How do we know if all this effort has really made any difference? If your company has a goal for adding new customers, a trailing indicator of service innovation would be growth. A leading indicator of service innovation would be customer reaction at the point of service. These reactions can be captured by asking the frontline employees for feedback. Customer feedback surveys are also a great tool, but lack immediacy. Ultimately, you know if this is working by how many new customers you gain as more people brag about their experiences. This can be captured by monitoring social media, and there are many tools for getting alerts.

Companies have even engaged with followers and have opened the door to have followers start doing the heavy lifting of marketing themselves. By "shouting out" to their fan base, companies invite brand advocates to host parties and provide care packages. They know their loyal fans want to share the outstanding experiences with their friends, and people are actually thrilled to help!

As the word spreads, you get more brand advocates who will share their experiences on the many social networking mediums. Your employees interact with many other people outside of the work environment, and you want them to become brand ambassadors of your value proposition. As the work environment improves, your employees are more likely to be spreading that word to people they know within their communities. All this will help increase top-line growth, and ultimately, that's how service excellence will be measured.

Going Forward

The key to Gemba walks for service innovation is to create an environment where people feel their ideas are valued. By opening a space for people to become involved and engaged in how the work is designed demonstrates respect. It also opens the door to humble appreciative leadership. The ability of a leader to say, "I don't know all the answers" opens the door wider to engagement. Leaders earn trust from their people by asking for help. It's through the helping relationship that we learn to trust each other. Gemba walks are a great way for people to learn keen observation skills, and by looking at other service providers we can find new customer delighters. By studying organizations that are outside of your own industry, you increase the chance to leapfrog your competition and generate many new service delighter ideas. Most importantly, a delighter needs only a slight presence for it to cause a huge increase in customer satisfaction. This means a very modest investment can result in a great deal of customer delight. Delighting customers is what this book is really all about.

Of the many key insights in this book, the centerpiece is understanding the human performance system. Please keep this in mind in whatever you do.

Many organizations look at continuous improvement as being small, incremental steps, with nothing earth shattering. Using Gemba walks to identify new service delighters can result in step change or discontinuous improvement.

About the Author

Robert (Bob) Petruska is an independent consultant who helps organizations identify and remove barriers that prevent employees from delivering an exceptional customer experience. He has experience in the aerospace, automotive, healthcare, food, office, and service industries. Bob likes to take teams on a fun-filled journey where passion for improvement meets innovative best practices. Teams take pride in creating their own success blueprints, and then rapidly  implement positive changes to realize their desired future states. Organizational energy spreads as age-old issues are quickly addressed, and new opportunities emerge as more and more people join in the fun.

Bob is an avid presenter whose energy and enthusiasm is contagious, as evident by positive participant feedback. He also holds a Bachelor of Science in industrial technology and a Master of Science in manufacturing systems from Southern Illinois University. Bob is a senior member of the American Society for Quality (ASQ), and is a Certified Six Sigma Black Belt.

Index